careers in

PRINTING

Mary Davis

photographs by
Milton J. Blumenfeld

Lerner Publications Company
Minneapolis, Minnesota

LIBRARY OF CONGRESS CATALOGING IN PUBLICATION DATA

Davis, Mary.
Careers in printing.

(An Early Career Book)
SUMMARY: Brief descriptions of the varied careers available in printing including those of linotype operator, proofreader, cameraman, press feeder, folder operator, and others.

1. Printing as a trade—Juvenile literature. [1. Printing as a trade] I. Blumenfeld, Milton J., illus. II. Title.

Z243.A2D33 686.2′023 72-5416
ISBN 0-8225-0306-9

 Published simultaneously in Canada by J. M. Dent & Sons Ltd., Don Mills, Ontario.

International Standard Book Number: 0-8225-0306-9 Library of Congress Catalog Card Number: 72-5416

Would you like to work in the printing business?

Printing is a very important business in the United States. Look around you and count all the things that have been printed. In almost every city you will find many printers—some large, some small.

Most people who work in printing start out as *apprentices*. This means that they earn money while they are learning their job.

There are many different jobs in the printing field. In this book you will learn about some of them.

PRINTING SALESMAN

The printing salesman has a very important job. He works with the customers. When a customer wants something printed, he talks to the salesman. The printing salesman may show him samples of the colored inks and the kinds of type used in printing. Then the salesman takes the customer's *order*—he writes down exactly what the customer wants. Customers depend on the printing salesman to make sure that their printing is done the way they want it.

The printing salesman tries to be very helpful because he wants people to do business with his company.

PRODUCTION MANAGER

The production manager is in charge of the print shop. He plans the printing jobs and knows the best way to handle a customer's order. The production manager must make sure that everyone is doing his work. He also has to make sure that the printing job will be finished when the customer wants it.

The production manager probably worked at many jobs in the printing company before he became a manager. He knows how all the equipment works and what everyone does in a printing company.

ARTIST

The artist plans the way a printing job will look when it is finished. She prepares a *layout*, or model, of each printing job. The layout shows the customer exactly how the printing job will look.

In some printing companies the artist may also make the *keyline*. She pastes the type for each page of the printing job onto a piece of heavy white cardboard. She also draws squares showing where pictures will go. Later on, you will find out how important the keyline is in the printing business.

The artist in a printing company has a lot of good ideas. Customers depend on her to make sure that the finished printing job is nice to look at.

1973
LERNER
books
1973
LERNER
books
corporate
corporate
image
Dominance
Dominance Bold
computer system

LINOTYPE OPERATOR

The type for a printing job can be made by different kinds of machines. A linotype operator uses a machine something like a typewriter. But it is much larger and much harder to use. As the operator types on the keyboard, the linotype machine makes lines of metal letters. The letters form words and sentences. When the lines of type are coated with ink, they can be transferred to a piece of paper called a *proof*.

The linotype machine gets its name from the words "line of type." It takes a lot of practice to run this complicated machine.

PHOTOTYPE OPERATOR

Remember how the linotype operator uses metal to make words? The phototype operator makes words on film. He types the words on a special kind of type-writer that makes a coded tape. The tape is put into a machine that has a small computer and a camera inside it. This machine produces a piece of special paper with a picture of the words and sentences on it.

Phototype operators have to be good typists. They must be fast, and they should not make any mistakes.

PROOFREADER

A proofreader checks all the type set by the print shop. She must check it very carefully. Otherwise there might be mistakes in the finished printing job. When the proofreader finds a mistake, she circles it with a red pencil. Then she corrects the mistake in the margin of the paper.

Are you a good speller? A proofreader has to be. A good profreader would know which word in this sentence is misspelled.

CAMERAMAN

Remember the keyline that the artist made? The cameraman takes a picture of the keyline, using a special kind of film. He also takes a picture of the *illustrations* —the photographs or drawings—to be used in a printing job. The cameraman sets his camera so that the pictures come out just right—not too dark or too light, not too big or too small. After the cameraman has developed his film, he checks it to make sure everything is just right. If there are mistakes, he must take another picture.

Many people take pictures just for fun. But taking pictures is a very serious job for a cameraman in a print shop.

ROBERT

LITHO ARTIST

After the cameraman's film is developed, it goes to the litho artist. The litho artist arranges the pieces of film on a large sheet of paper called a *flat*. He arranges the film with the words and pictures on it exactly as the artist arranged the words and pictures on the keyline.

The litho artist is also called a *stripper*, and his work is called *stripping*. The stripper must do his work very carefully, without making mistakes.

PLATE MAKER

The plate maker is the last person to handle the printing job before it is put on the printing press. He prepares the plates for the press. A *press plate* is a thin piece of metal that is put into the printing press. The plate maker takes the litho artist's flat and puts it in front of the metal plate. By using a very bright light, he transfers the words and pictures from the film to the plate.

The plate maker must work in a very clean room. Pieces of dust or dirt can spoil the plates.

INK MATCHER

Sometimes colors are used in printing. It is very important that the colors used in a printing job look the way the customer wants them to look. It is the job of the ink matcher to make sure that the colors please the customer. Sometimes the ink matcher has to mix several inks together to get the color he wants.

In this picture you see *ink draw-downs*, samples of color that the ink matcher shows to the customer.

CS-310
CS-200
CS-100
CAUTION

PRESSMAN

The pressman is in charge of running the printing press. He watches the press very carefully. Sometimes there is too much ink on the press. This can make the printed words look too dark. When there are problems, the pressman stops the press.

The pressman is a good mechanic. If the press breaks down, he knows how to fix it.

PRESS FEEDER

It is the job of the press feeder to put paper into the printing press. He has to make sure that the paper is going in the right way. He also has to have enough paper ready to go into the press. He helps the pressman in other ways too.

The press feeder is often an apprentice in the printing plant. After he knows his job, he can learn to become a pressman.

PRESS FOREMAN

The press foreman looks at the first sheets of paper that come off the printing press. In this picture he is using a magnifying glass to look at a sheet of paper. He makes sure that the colors are right, and that the words and pictures are printed straight. If something is wrong, he helps the pressman to correct the problem.

The press foreman is very particular. He wants the printing job to be exactly right.

FOLDER OPERATOR

The last place that the printing job goes is the *bindery*. Here the printed sheets are folded and put together. The folder operator runs the bindery's *folding machine*. This machine can fold printed sheets in many different ways. The folder operator has to follow instructions carefully for each printing job.

The folder operator often has helpers. In this picture, one of his helpers is taking finished printing from the folding machine and packing it for the customer.

DELIVERYMAN

The deliveryman takes the finished printing to the customer. Sometimes he takes it in a car. Sometimes he uses a truck. When the packages of printing are very heavy, the deliveryman has to use special equipment to move them.

Like the salesman, the deliveryman meets many people. He tries to be very helpful at all times.

2013
ON AVENUE NO.
MINN. 55401

Printing careers described in this book

Printing Salesman

Production Manager

Artist

Linotype Operator

Phototype Operator

Proofreader

Cameraman

Litho Artist

Plate Maker

Ink Matcher

Pressman

Press Feeder

Press Foreman

Folder Operator

Deliveryman

A letter from a printing company executive

300 FIRST AVENUE NORTH / MINNEAPOLIS, MINNESOTA 55401

Dear Readers,

After reading this book we think you will agree that printing is an interesting and exciting business.

These pictures show you that there are many different jobs in printing. The next time you have a chance to visit a printing plant we are sure you will enjoy seeing the people who are working at these different jobs.

And when you grow up, maybe you will decide that not only will it be rewarding, but it will also be fun to work in the printing industry.

Sincerely,

David Z. Johnson, President

The publisher would like to thank Handschy Chemical Company, Johnson Printing Company, Inc., Meyers Printing Company, P & H Photo Composition, and True Color Corporation for their cooperation in the preparation of this book.